I0879034

Disaster Zone

The Science Behind SUPERVOLCANOES

Louise Spilsbury

CHERITON
CHILDREN'S BOOKS

Please visit our website, www.cheritonchildrensbooks.com to see more of our high-quality books.

First Edition

Published in 2022 by **Cheriton Children's Books**
PO Box 7258, Bridgnorth WV16 9ET, UK

Author: Louise Spilsbury
Designer: Paul Myerscough
Editor: Tim Cooke
Proofreader: Wendy Scavuzzo
Picture Researcher: Rachel Blount
Consultant: David Hawksett, BSc

Picture credits: Cover: Shutterstock/Roberto Destarac Photo; Inside: p1: Shutterstock/Fboudrias; pp4-5: Shutterstock/Wead; p4b: Shutterstock/www.sandatlas.org; p6: Wikimedia Commons/SGT PAUL BISHOP; p7: Wikimedia Commons/יקד; p8: Wikimedia Commons/Neureiter; p9: Shutterstock/Balounm; pp10-11: Wikimedia Commons/Georesearch Volcanedo Germany; p11r: Wikimedia Commons/NASA Expedition 20 crew; p12-13: Shutterstock/Budkov Denis; p13r: Shutterstock/Peter Hermes Furian; p14: Shutterstock/Ronn Hook; p15: Shutterstock/Gnoparus; p16: Shutterstock/Fboudrias; p17: Shutterstock/Budkov Denis; pp18-19: Shutterstock/Valeriy Poltorak; p19r: Wikimedia Commons/Danial Dzurisin; pp20-21: NARA/Gempis, Val/TECH. SGT.; p21b: Wikimedia Commons/Gringer; p22: Shutterstock/Photolike; p23: Shutterstock/Aleksei Potov; pp24-25: Wikimedia Commons/USGS; p25b: Shutterstock/Benny Marty; pp26-27: Wikimedia Commons/USGS; p27b: Wikimedia Commons/W. Chadwick; p28: USGS Hawaiian Volcano Observatory (HVO); p29: NASA/GSFC/METI/ERSDAC/JAROS, and U.S./Japan ASTER Science Team; p30: Shutterstock/Fishman64; p31: US Geological Survey, Washington State University, Jet Propulsion Laboratory, California Institute of Technology; pp32-33: Wikimedia Commons/Dr. Carlos Costales Terán; p33r: Shutterstock/Jo Reason; pp34-35: Shutterstock/Josemaria Toscano; p34b: Shutterstock/Photovolcanica.com; p36: USGS; p37: Wikimedia Commons/Jebulon; pp38-39: Shutterstock/Kris Wiktor; p39r: Shutterstock/Jody Ann; pp40-41: Shutterstock/Marco Ossino; pp40-41b: Shutterstock/Michael W NZ; p42: Shutterstock/Anglea N Perryman; p43: Shutterstock/SpeedShutter; p44: Shutterstock/Fotosdelalma; p45: Wikimedia Commons/ZiaLater; p47: Shutterstock/www.sandatlas.org.

Printed in the United States of America

Contents

Chapter 1

ERUPTIONS OF EARTH

A volcanic eruption is one of the most amazing but scary **forces of nature**. Volcanoes are vents, or openings, in Earth's surface. They can remain peaceful for hundreds or even thousands of years. However, when the **pressure** below Earth's surface builds up, volcanoes can erupt, or explode. Volcanic ash, gases, and fiery-hot liquid rock can be blasted into the sky and over the surrounding area. Such eruptions can cause destruction, damage, and death far and wide.

Super Eruptions

Around the world, volcanoes are erupting all the time. Most of these eruptions are small, for example, a small volcano may splutter out just a slight burst of steam and ash. Such an eruption may cause a minor annoyance, for example, it might leave a thin layer of ash and dust on cars and homes that people have to wash off. People should fear larger volcanic eruptions. The scariest are the rare, but enormous, super eruptions, which send out huge amounts of rock. They release violent blasts of burning ash, gas, and lava, which is hot, liquid rock. Eruptions from these supervolcanoes can affect the whole world.

Lava fresh from underground is nearly three times as hot as a pizza oven.

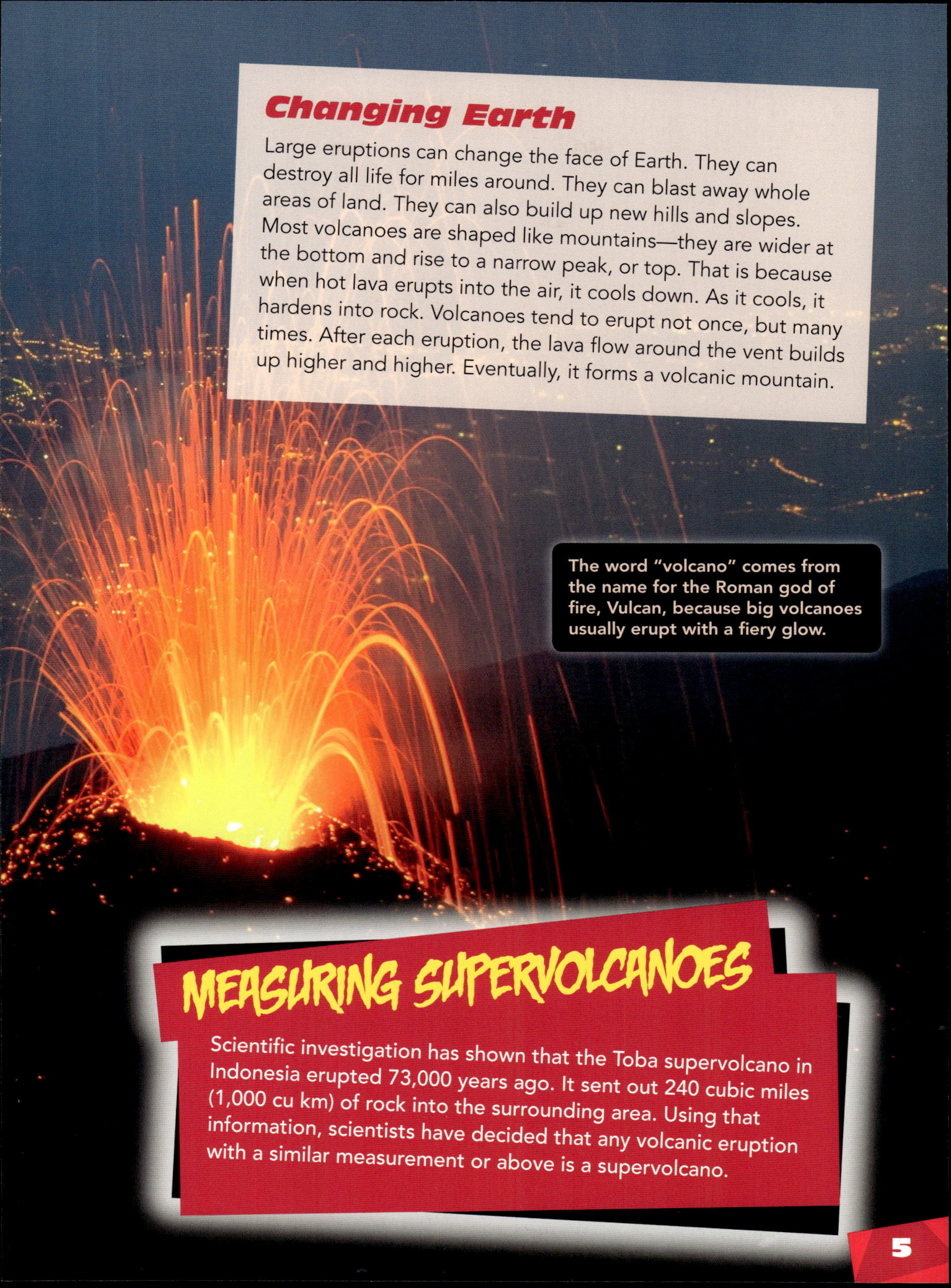

Changing Earth

Large eruptions can change the face of Earth. They can destroy all life for miles around. They can blast away whole areas of land. They can also build up new hills and slopes. Most volcanoes are shaped like mountains—they are wider at the bottom and rise to a narrow peak, or top. That is because when hot lava erupts into the air, it cools down. As it cools, it hardens into rock. Volcanoes tend to erupt not once, but many times. After each eruption, the lava flow around the vent builds up higher and higher. Eventually, it forms a volcanic mountain.

The word "volcano" comes from the name for the Roman god of fire, Vulcan, because big volcanoes usually erupt with a fiery glow.

MEASURING SUPERVOLCANOES

Scientific investigation has shown that the Toba supervolcano in Indonesia erupted 73,000 years ago. It sent out 240 cubic miles (1,000 cu km) of rock into the surrounding area. Using that information, scientists have decided that any volcanic eruption with a similar measurement or above is a supervolcano.

DEADLY DISASTERS

The ash, rock, and lava that volcanoes spit out can cover huge areas of land and destroy whole towns and villages in their path. Some volcanoes send a river of burning-hot lava down their sloping sides. If the lava is very thick and sticky, it may move slowly enough for people to run or drive out of its path. Other types of lava can flow as fast as a speeding car—up to 124 miles per hour (200 kph). They can spread long distances over the land surrounding the volcano. All lava is heavy and hot. As it flows along, it crushes objects and sets fire to trees, buildings, and anything else that is in its way.

Ash Attacks

The ash that blasts out of a volcano during an eruption is made up of tiny **particles** of rock. They are so small that they look like dust. Ash-filled air can choke people and can cause road and railroad accidents. The ash can also be blown by wind and suddenly fall to cover fields, rivers, and towns. In 1991, millions of tons of ash were blasted into the air by Mount Pinatubo in the Philippines. Strong winds blew the ash in all directions, including into a severe storm called a typhoon. It wet the ash and made it very heavy. It fell in thick layers onto buildings. Many roofs collapsed, killing 300 people inside their homes.

The explosive eruption of Mount Pinatubo in the Philippines in 1991 sent out massive ash clouds that fell and smothered huge areas of land.

Volcanic Bombs

Powerful eruptions can also shoot out chunks of rock. Any piece of flying volcanic rock bigger than about 2.5 inches (6.4 cm) in diameter is known as a volcanic bomb. Volcanic bombs form from bits of lava that are blasted into the sky. They cool and harden quickly in the air before they hit the ground. Some volcanic bombs are huge. They can be tossed long distances away from the volcano.

Where in the World?

The Galeras volcano in Colombia erupted unexpectedly in 1993. Giant lumps of volcanic matter, called volcanic bombs, flew through the air. When Galeras erupted, six volcanologists, or scientists who study volcanoes, and three tourists were killed near the **summit**. They were hit and killed by the bombs and several others were seriously injured by the blasted rock.

This lava bomb has cooled and hardened. Cracks formed on the surface of the bomb as it cooled.

A TRAIL OF DESTRUCTION

Large volcanoes are so dangerous that they can devastate cities and can kill a lot of people. Some of the most violent volcanoes also set off other natural, but equally deadly, disasters that cause further destruction.

Lethal Lahars

A lahar, or mudflow, happens when the heat from an eruption melts snow and ice on a volcanic mountain. The melted snow and ice creates liquid water. It mixes with ash and rock pieces as it flows down the slopes of the volcano. It pours down a mountainside like wet concrete. As it speeds along, it picks up bits of rock, soil, trees—it can even carry along buildings and bridges. It becomes bigger and bigger and more and more dangerous. Large mudflows can crush, bury, or carry away almost anything in their paths.

Pyroclastic Flows

Pyroclastic flows are perhaps the deadliest of all volcanic dangers. They are mixtures of **superheated** gases, ash, and rock. They can speed down slopes at up to 450 miles per hour (724 kph). They look like an **avalanche** of ash and their temperatures can be more than 1,300 degrees Fahrenheit (704 °C). Pyroclastic flows can travel long distances and are impossible to outrun. Once they are near a town or village, the deadly waves of heat and gas will boil and burn objects and people instantly.

In 1991, volcanic ash washed down the slopes of Mount Pinatubo in an enormous lahar, destroying everything in its path.

Where in the World?

In 79 AD, Mount Vesuvius erupted in Italy. The pyroclastic flow hit the Roman cities of Pompeii and Herculaneum. Many people had managed to escape at the first signs of eruption. However, thousands of other people were killed. The hollow shapes of their dead bodies were **preserved** in the ash.

The bodies of the people killed in the Mount Vesuvius disaster are preserved in the positions in which they died.

Avalanches and Tsunamis

Sometimes, heat from lava or pyroclastic flows melts vast chunks of ice from **glaciers** or snow fields. The ice breaks off and slides down slopes in deadly avalanches. The sides of a volcano can collapse into surrounding ocean during an eruption, and a giant pyroclastic flow may also crash into the water. Those impacts may cause giant waves called tsunamis. Eruptions from underwater volcanoes can cause tsunamis, too. As tsunamis near shores, the waves become taller and taller until they hit the coastline with devastating force. They cause mass destruction as they smash into coastlines.

Q

The preserved bodies in Pompeii show that many people were trying to cover their faces when they died. Why do you think they did so?

Disaster Zone

Mount Tambora, Indonesia

In 1815, one of the largest and deadliest volcanic eruptions in recorded history took place. It was at Mount Tambora on the island of Sumbawa, Indonesia. When this mega monster blew its top, it destroyed the surrounding area and killed 10,000 people. It also destroyed 35,000 homes.

Explosion!

The series of large explosions at Mount Tambora shot huge amounts of gases, dust, **pumice**, and other rock into the air and the surrounding area. Some of the rocks were as big as small cars, however, they were tossed into the air like feathers. Wide streams of fiery ash and rock raced down the mountain's sides at more than 100 miles per hour (160 kph). They burned grasslands and forests. When the pyroclastic flow of gas, ash, and rock reached the cold ocean, it formed giant, floating lumps of pumice rock that trapped ships in the harbor.

TRIGGERING A TSUNAMI

When the pyroclastic flow from Tambora reached the ocean, it entered the water with great force. It triggered tsunami waves more than 16 feet (5 m) high. The waves spread out from the island of Sumbawa and across the Java Sea like ripples on a pond. The waves smashed into other Indonesian islands nearby and caused **flooding**, devastation, and death.

The Year Without Summer

Mount Tambora sent at least 60 megatons (54 MT) of **sulfur** into the **atmosphere**. These particles and other tiny bits of **debris** from the volcano floated in the air. They drifted on the wind over a wide area, blocking out a lot of the sun's heat. The average global temperature fell as much as 5.4 degrees Fahrenheit (3 °C). The effect was so great that 1816 became known as the "year without a summer." The change in temperature killed plants and **crops**. It led to people starving in parts of North America and Europe. The eruption claimed the lives of thousands of people around the world. Its final death toll was around 80,000.

Tambora is a volcanic mountain that forms one of the highest peaks in Indonesia. It is still **active** today.

Before it erupted, Mount Tambora was almost 14,000 feet (4,267 m) high. After the center of the volcano collapsed, it left a **crater** 3.7 miles (6 km) wide.

Q How do you think people today would respond to an enormous volcanic disaster, like that of Mount Tambora? Are we better able to cope, or would the impact of the disaster be worse today? Give reasons for your answer.

Chapter 2

HOW VOLCANOES WORK

A volcanic eruption happens when a volcano vent is forced open. Red-hot liquid rock, called magma, as well as gases and pieces of rock from inside Earth, burst out. From the smallest to the biggest and deadliest volcanoes, eruptions start deep within the layers of Earth.

Inside Earth

To understand volcanoes, we have to look deep into Earth to see beneath the surface. Earth is made up of different layers. The outer layer that we live on is called the crust. It is built from rock that forms the land and the floor of the world's oceans. It is between 4 to 50 miles (6 to 81 km) deep. The crust is the thinnest of Earth's layers. Around 3 billion years ago, the crust cracked and split into sections that fit over the surface of Earth like the panels on a soccer ball. These huge rafts of rock are called tectonic plates.

On the Move

Although we cannot feel them moving, Earth's tectonic plates are constantly, but very slowly, shifting. They float on the upper layer of the mantle. That is the layer below the crust and it is the thickest of Earth's layers. The mantle is made up of melted and partly melted rock, **dissolved** hot gases, and pieces of solid rock. The mantle is heated by the center of Earth, which is called the core. Magma usually stays deep underground, but because it is lighter than the surrounding solid rock, it sometimes rises up. Then it squeezes out through gaps between the plates on the crust. Volcanoes are the places where magma makes it through to the surface of Earth.

HEAT FROM THE CORE

At the center of the planet is Earth's inner core. It is a superheated ball of metal that can reach temperatures of 9,900 degrees Fahrenheit (5,482 °C). The layer outside the inner core is called the outer core, and it is made of **molten** metal. As the core heats up the mantle, the magma constantly moves because the heat makes it rise. As it rises nearer the surface, it cools, and as it cools, it sinks. That rising and falling of magma makes the tectonic plates on top of the mantle move.

Once magma erupts from a volcano, it is called lava. It can range from 1,300 to 2,400 degrees Fahrenheit (704 to 1,316 °C).

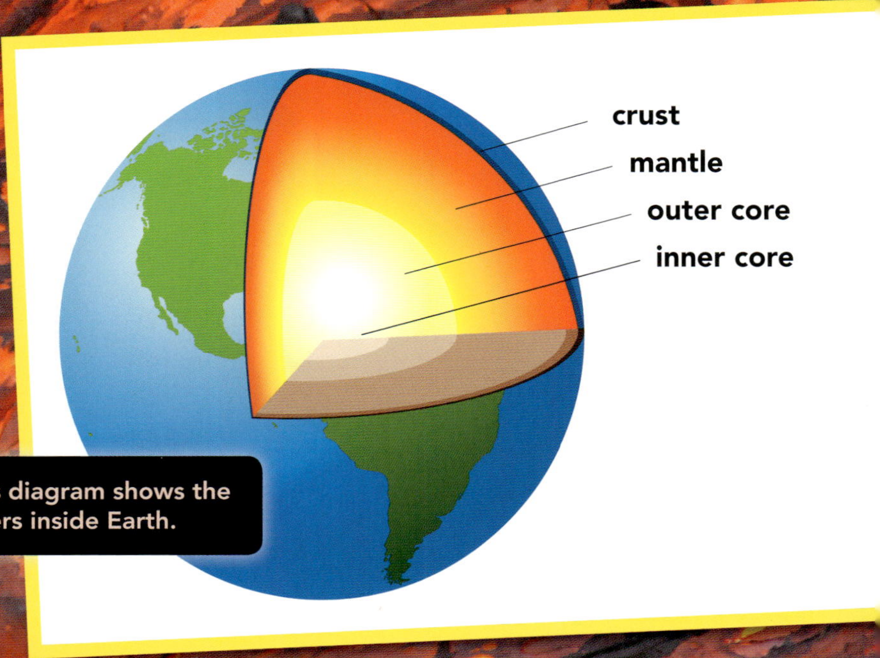

This diagram shows the layers inside Earth.

UNDER PRESSURE

Volcanic eruptions happen after hot magma rises near the surface of Earth's crust. Rising magma reaches closer to the surface by forcing its way through cracks in the crust. Magma is super-hot and can melt solid rock above it. That creates a tunnel through which magma can then move upward.

Magma Chambers

Near the upper part of the crust, just below Earth's surface, it is cooler. There, the magma becomes cooler and thicker, and moves more slowly and less powerfully. It becomes stuck near the surface. Gradually, more magma rises and joins the magma already stuck at the surface. Over time, a large area of magma builds up. That is known as a magma chamber. The chamber of trapped magma is pushed slowly upward by more rising magma from below. When it pushes against weak rock above it, cracks may appear. Alternatively, the ground may bulge a little because of the pressure build-up from below.

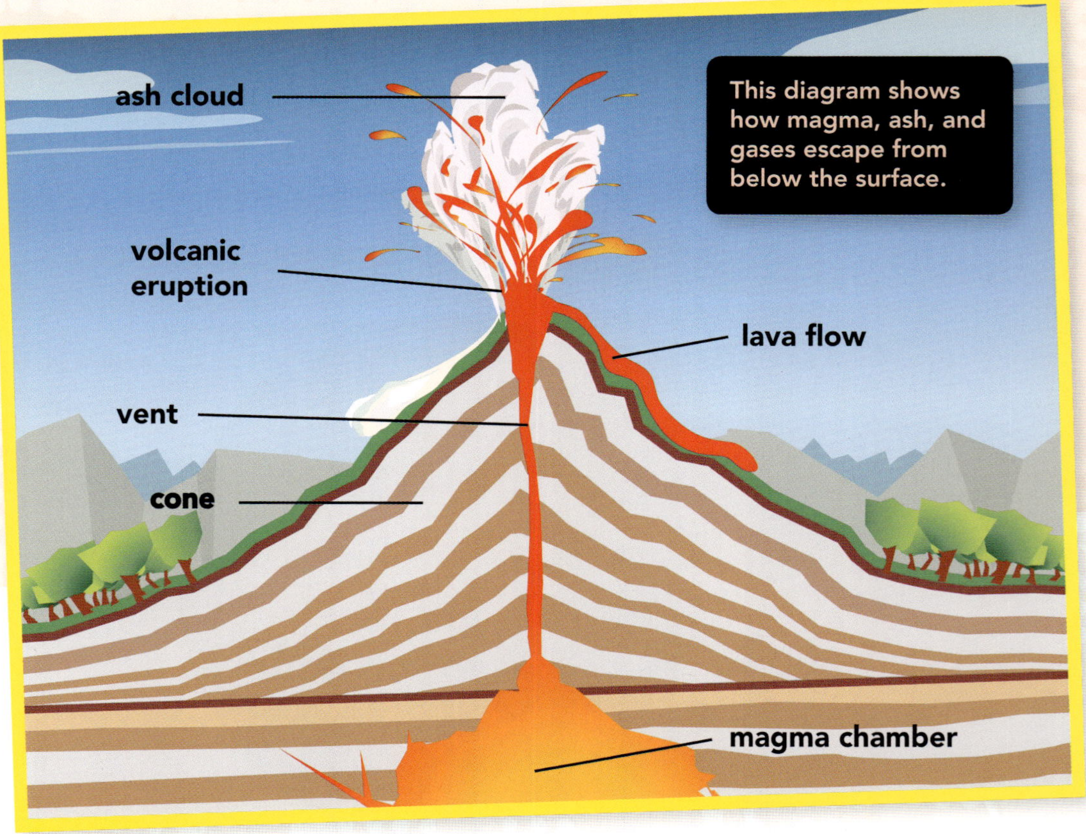

This diagram shows how magma, ash, and gases escape from below the surface.

As a volcanic eruption happens, rock and other materials explode outward and away from the vent. That explosion leaves a crater.

Too Much Pressure

When the pressure of built-up magma finally becomes too much, the magma blasts through the crust. That is a volcanic eruption. The hole in the center where eruptions usually happen is called the vent. The magma escapes through a vent at the surface into the air. Now it is called lava. Lava can burst out of the top of a volcano like soda from a shaken bottle. It can also seep through holes in the slopes of a volcanic mountain. That happens when magma melts horizontal tunnels through rock and, instead of traveling upward, flows sideways.

BIG AND SMALL

Small volcanoes erupt more often than bigger volcanoes. It takes a very long time for large amounts of magma to create a big volcano, which leads to a large eruption. Some smaller volcanoes erupt quietly every few months or years. There is never enough magma build-up below them to cause a big or super eruption.

EXPLODING EARTH

The most violent and dangerous eruptions happen when volcanoes seem to explode. They shoot lava, ash, and gas high into the air. How explosive a volcano is depends on the type of magma inside it, which can be either runny or sticky as pressure builds up on it before an eruption.

Different Types of Magma

Thin and runny magma contains less gas because gases can easily escape from it. When that type of magma erupts, it flows out of the volcano and down its sides. That is known as an effusive eruption. The lava usually moves slowly enough for people to move out of the way to safety. Gases cannot escape easily from thick and sticky magma, and the pressure caused by the bubbles of gas builds up, until the gas suddenly escapes and explodes. Then the magma blasts into the air and becomes lava. As the lava cools in the air, it breaks into pieces that vary in size from tiny bits of ash to huge boulders. That is known as an explosive eruption.

Mount Fuego in Guatemala is one of the world's most active volcanoes. This photograph shows an explosive eruption, which took place in June 2018.

The stickiness of lava depends on the chemicals and minerals that it contains. For example, the stickiest lava usually contains a lot of **silica**.

Fire Fountains

A fire fountain is an explosive eruption that keeps going, releasing a continuous spray of lava. The lava shoots into the sky like a fountain, and its sprays can keep flaring into the air for days on end, lighting up the night sky. Some fire fountains can reach hundreds of feet above the volcano vent! Some supervolcanoes create eruption columns that are more like fountains of ash. They happen when the force of a blast is so great that it can turn magma into tall, vertical clouds of ash that can travel up to 35 miles (56 km) into the sky.

Where in the World?

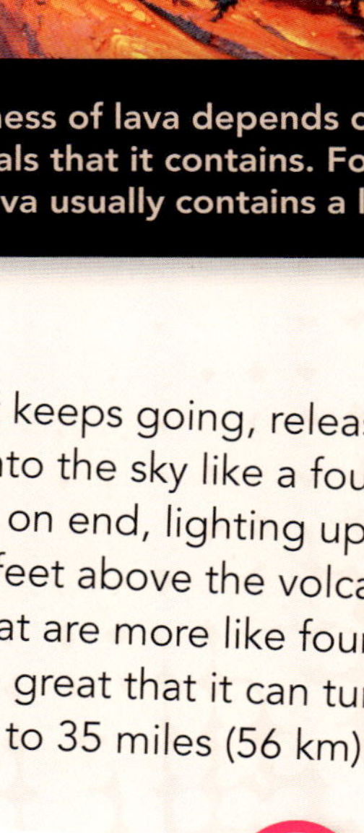

Some effusive eruptions carry on for a long time and can produce a lot of lava. The Plosky Tolbachik volcano in Kamchatka, Russia, started to erupt on July 6, 1975. It did not stop until December 10, 1976. In 2012, the same volcano erupted again. It poured out around 1,200 tons (1,090 mt) of lava every second.

Disaster Zone

Mount St. Helens, Washington

At 8:32 a.m. on May 18, 1980, the Mount St. Helens volcano, among the Cascade Mountains in the state of Washington, erupted. It became the worst volcanic disaster ever to happen in modern North America. That explosive eruption transformed a beautiful, forest-covered mountain into a gray, empty wilderness.

Mount St. Helens' cone was blasted away by the 1980 eruption, leaving a crater 1.2 miles (1.9 km) wide and 1.8 miles (2.9 km) long.

A Deadly Eruption in Action

When Mount St. Helens erupted, a huge section of the north side of the mountain was blasted off. It started a massive **landslide** that rolled down the slope and suffocated a wide area of land below. The eruption spat out a column of gas and ash 16 miles (26 km) high that blew through the air. The ash fell as far away as central Montana and brought total darkness to Spokane, Washington, about 250 miles (402 km) away. Volcanic gases exploded out of the side of the volcano. They traveled at high speeds, and spread up to 19 miles (31 km) away. The eruption also caused mudflows, pyroclastic flows, and floods that buried anything in their path. Even valleys as far as 17 miles (27 km) away were covered in deep mud and debris.

After the Disaster

When the dust finally settled, it was obvious that a huge area around the volcano had been destroyed. The blast had devastated everything in its path up to about 8 miles (13 km) away and a total area of about 229 square miles (596 sq km) was destroyed. Fifty-seven people and thousands of animals, including deer, elk, and bears, were killed. Trees were flattened between 8 miles (13 km) and about 13 miles (21 km) away from the blast. They were knocked down as if they were tooth picks. It cost more than $1 billion to clean up the area. Scientists today are closely watching the Mount St. Helens volcano to be prepared for future explosions.

A photographer was killed when his car was surrounded and filled with ash from the volcano.

WARNING SIGNS

There were warning signs that Mount St. Helens was going to blow. From about March 16, there was a series of small **earthquakes** in the area that only scientists noticed. Then on March 20, a larger earthquake shook the region. A continuous shaking of the ground known as a volcano tremor warned scientists that magma build-up and movement were taking place. In the five weeks before its eruption, the top of Mount St. Helens bulged out by 450 feet (137 m). That was because of the trapped magma building up underground.

Q

How do you think people can be prepared for the next Mount St. Helen's explosion? What measures would you put in place if you lived nearby?

Chapter 3

WHERE DO VOLCANOES OCCUR?

Volcanoes, big and small, mainly happen at places where tectonic plates meet. That is called a plate boundary. At plate boundaries, there are natural cracks in the crust that magma can more easily force its way up and through. Volcanoes also occur in parts of the crust that lie over mantle **plumes**. Mantle plumes are areas of super-hot, fiery rock below Earth's surface.

The Ring of Fire

Around 450 volcanoes line the edges of the Pacific Ocean. That area is at such high risk of red-hot lava eruptions that it is called the Ring of Fire. The Ring of Fire runs north from the base of Chile up to Alaska. It runs south below Siberia and on to Japan, the Philippines, and finally New Zealand. The Ring of Fire is where the huge Pacific plate, which lies beneath the ocean, joins several other tectonic plates that lie beneath **continents**. The areas where the continental plates meet the Pacific **oceanic plate** are mostly convergent boundaries. That means that the plates there move toward each other, and push against or slide over each other.

Volcano Factory

There are a lot of volcanoes along the Ring of Fire. The Pacific plate is jammed beneath the continental plates. Oceanic plates are heavier than continental plates because they are made up of heavier metals. As the edge of the Pacific plate dips down into the mantle, some of the water within it becomes hot and boils, which creates steam. That superheated steam helps produce more magma at the boundary so more volcanoes form there.

Where in the World?

When Mount Pinatubo erupted in 1991, it blasted an enormous cloud of ash into the air that was 22 miles (35 km) high and 300 miles (483 km) wide. The ash immediately turned day to night around the volcano. Three weeks later, the ash had spread around the world. It blocked sunlight and made temperatures drop by 1 degree Fahrenheit (0.6 ^{0}C). The eruption of the volcano killed about 300 people and made another 100,000 homeless.

The ash cloud created by the 1991 Mount Pinatubo explosion can be seen in this photograph.

In total, the horseshoe-shaped Ring of Fire measures nearly 25,000 miles (40,233 km) long.

BENEATH THE WATER

Volcanoes are noisy, violent acts of nature. People don't know about many of the volcanoes that erupt. That is because most volcanic eruptions happen deep underwater, on the floor of the oceans. They are usually hidden from view, covered beneath an average of 8,500 feet (2,590 m) of water.

Undersea Mountains

Underwater volcanoes often happen where tectonic plates meet on the ocean floor. These regions look like underwater **mountain ranges** and are called mid-ocean ridges. Mid-ocean ridges usually have a valley, known as a rift, running down the middle. They happen at boundaries where the plates are pulling apart and magma escapes to the surface. There, it cools to form rock, building up into peaks.

The Mid-Atlantic Ridge is a mostly underwater mountain range, but it can be seen in a spectacular way in Iceland.

Volcanic Islands

Many of the volcanoes that erupt along mid-ocean ridges are invisible at the surface. Large volcanoes that erupt more frequently gradually grow taller and taller. They behave as they do on land, to form mountains. In cool ocean water, lava can cool and harden into rock even more quickly than it does on land. Some underwater volcanoes grow so tall that their peaks stick out above the water to form islands in the sea. Some islands are mountains from which lava still sometimes erupts.

Hot Spots

Islands do not only form over a fault, or crack, where the edges of two oceanic plates meet. Some islands build up from volcanoes that occur over the center of an oceanic plate. That is at a point where a hot mantle plume burns through a slightly thinner part of the crust. That is called a hot spot. An oceanic plate moves along a few inches each year. However, the hot spot remains in the same place. Over millions of years, as the hot spot burns through the crust again and again, it can create a line of islands next to each other.

VOLCANOES IN THE BLUE

Scientists believe there could be more than 1 million underwater volcanoes around the world. About three-quarters of all the bubbling red-hot magma that escapes through Earth's crust does so along the mid-ocean ridges. They are deep below the surface of the ocean. In 2012, one of the world's largest volcanoes, Havre, exploded. It lies on the sea floor in the middle of the Pacific Ocean. A huge amount of lava poured out from 14 vents for about 90 days.

The Hawaiian Island chain in the Pacific Ocean was formed over a hot spot.

Q What areas of the world in particular do you think scientists should monitor underwater volcanoes, and why?

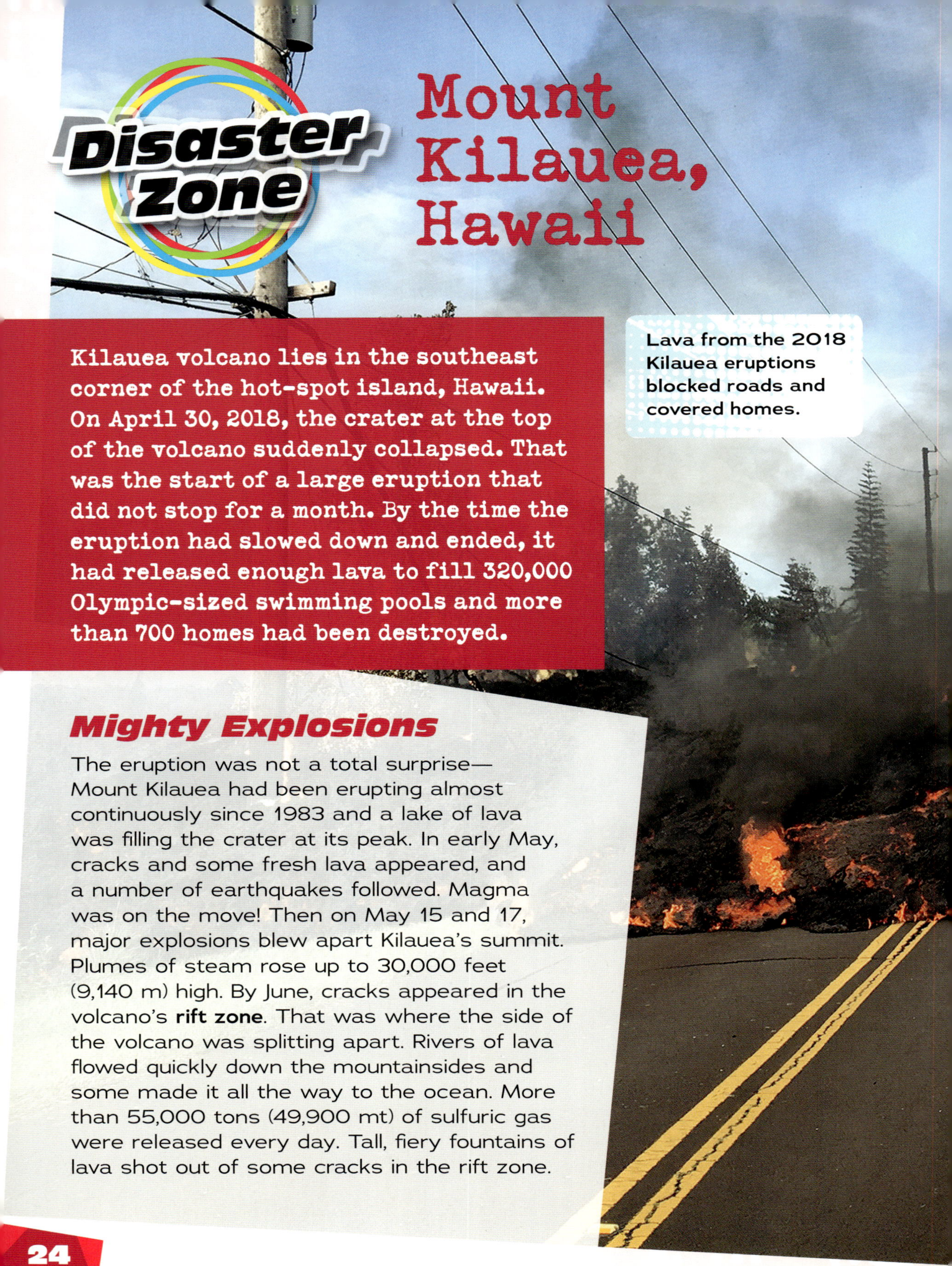

Lava from the 2018 Kilauea eruptions blocked roads and covered homes.

Disaster Zone

Mount Kilauea, Hawaii

Kilauea volcano lies in the southeast corner of the hot-spot island, Hawaii. On April 30, 2018, the crater at the top of the volcano suddenly collapsed. That was the start of a large eruption that did not stop for a month. By the time the eruption had slowed down and ended, it had released enough lava to fill 320,000 Olympic-sized swimming pools and more than 700 homes had been destroyed.

Mighty Explosions

The eruption was not a total surprise—Mount Kilauea had been erupting almost continuously since 1983 and a lake of lava was filling the crater at its peak. In early May, cracks and some fresh lava appeared, and a number of earthquakes followed. Magma was on the move! Then on May 15 and 17, major explosions blew apart Kilauea's summit. Plumes of steam rose up to 30,000 feet (9,140 m) high. By June, cracks appeared in the volcano's **rift zone**. That was where the side of the volcano was splitting apart. Rivers of lava flowed quickly down the mountainsides and some made it all the way to the ocean. More than 55,000 tons (49,900 mt) of sulfuric gas were released every day. Tall, fiery fountains of lava shot out of some cracks in the rift zone.

Changing the Island

By August 4, the eruptions stopped and 35 years of constant activity at the volcano ended. However, the eruptions had changed the island. About 13.7 square miles (35.5 sq km) of land were covered in lava. That is enough to bury the island of Manhattan about 27 feet (8 m) deep in lava! The summit of Mount Kilauea had collapsed by about 1,640 feet (500 m) and the island had grown outward. The spreading lava added an area of about 1.5 square miles (3.9 sq km) of land to Hawaii's total surface.

PREDICTING THE NEXT ONE

There is still magma beneath Kilauea's summit and rift zone. That means there will likely be another eruption on Kilauea in the future. Scientists are studying the 2018 eruption closely. They hope it can help them **predict** and deal with the next explosion.

As lava from Mount Kilauea streamed into ocean water, it formed new rock and land.

Chapter 4

SCIENTISTS ON THE CASE

Volcanoes cause death and destruction. Supervolcanoes have caused **climate** catastrophes and animal extinctions. Scientists are working to better understand how volcanoes work where they happen. They hope to predict when a volcano or a supervolcano might erupt in the future.

Volcanologists at Work

Scientists who study volcanoes are called volcanologists. They work in laboratories or research stations. They also work on-site at volcano locations. Volcanologists use rock hammers to chip off **samples** of rock from the area around volcanoes, then study them in a lab. Chemical and dating analyses of rock samples can, for example, reveal when past eruptions occurred and what they were like. It can show if there were pyroclastic flows, how large they were, and how far they spread. That type of information gives clues as to what future eruptions might be like.

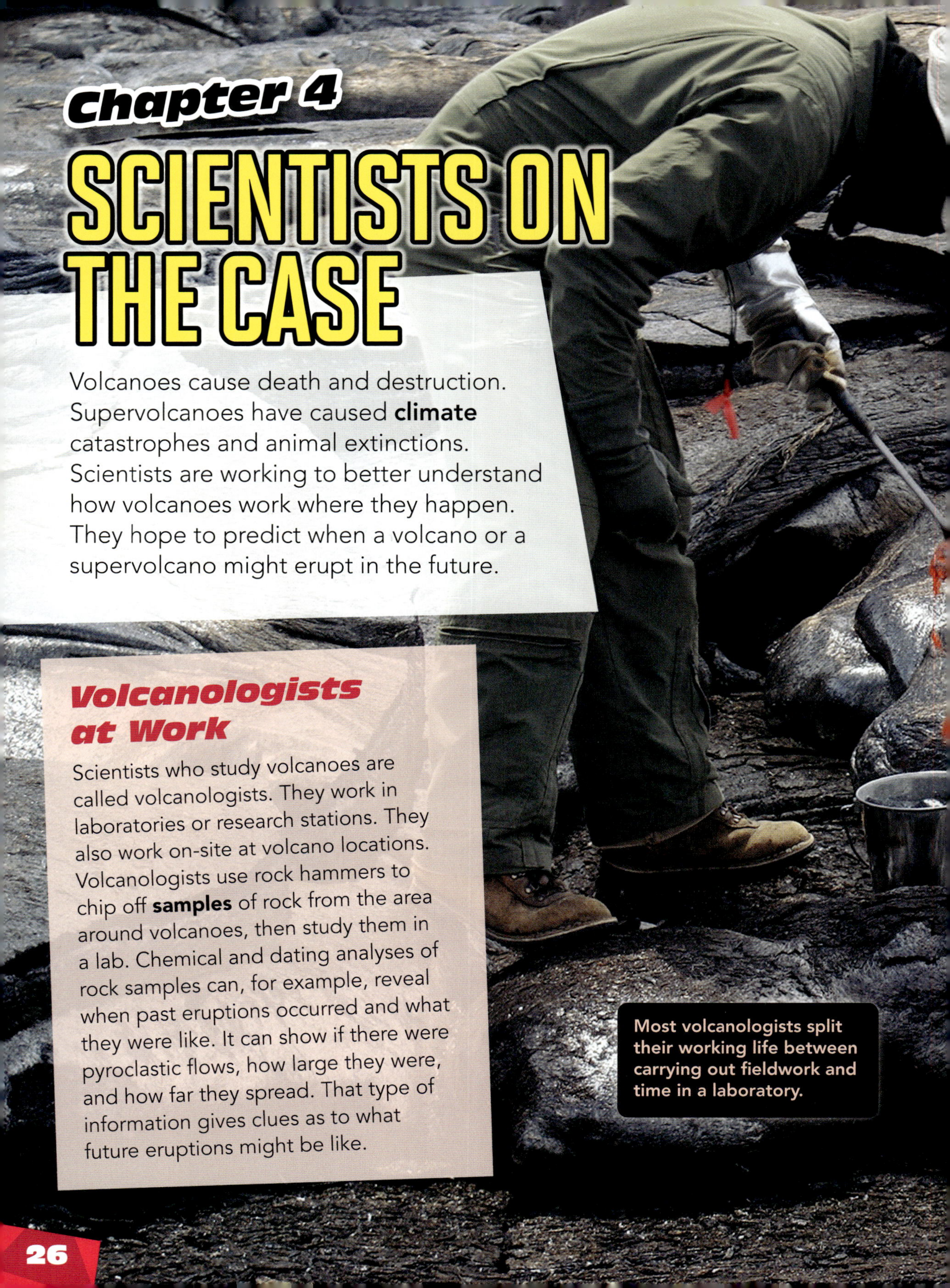

Most volcanologists split their working life between carrying out fieldwork and time in a laboratory.

Learning More

Volcanologists use many different tools and equipment to help them discover more about volcanoes. When magma rises through the crust toward Earth's surface, it can make the crust move. Devices called seismometers detect these vibrations in Earth's crust. They are called seismic waves. Seismic waves spread out from their starting point a little like ripples across a pond. They can travel through rock and can reach far from where the shock began. Seismometers often have a weight, which is held in place by springs. As the ground moves, the weight moves. That can tell volcanologists the size, direction, and speed of the seismic waves. That information is recorded on a graph by a machine.

MEASURING VOLCANOES

Volcanologists measure how big an eruption is by how much material, including magma, rock, and ash, it releases. The Volcano Explosivity Index (VEI) rates eruptions from 0 to 8, with 8 being the highest. Mount St. Helens' explosion in 1980 released 0.25 cubic miles (1 cu km) of material and was rated VEI 5. The Toba supervolcano released 240 cubic miles (1,000 cu km) of material, earning it the maximum VEI 8.

Volcanologists can use a tool called an electron microprobe to study small rock pieces to see what they are and how old they are.

VOLCANOES ON VIEW

It is almost impossible to predict volcanic eruptions. They are also very difficult to study close up. However, volcanologists have different ways of examining and watching volcanoes that they know could erupt at some point.

Seeing Heat

Thermal imaging cameras produce an image using the heat an object gives off. They can also use the heat to find an object. The camera can tell us the hottest or coldest parts of an object. Scientists point hand-held thermal cameras at an erupting volcano, or a volcano they think is about to erupt. The camera will record the heat level and where the heat is. That is useful because temperatures around the volcano rise as there is more magma activity below the surface. Thermal images also help scientists find places on a volcano where it is cool enough to put instruments to measure how much the ground moves.

This scientist is pointing a thermal imaging camera at a gap in the side of a volcano to take a reading.

Secrets from Satellites

Some instruments that view volcanoes are not even on Earth. The instruments are on **satellites** that **orbit** the planet high up in space. They can carry cameras that can see ash plumes from **remote** volcanoes. Some satellites can take images that show the temperature of the air in the atmosphere. That helps the experts because air often gets warmer above volcanoes that are about to erupt. Scientists also use satellites to see whether an area of ground above a volcano is moving or changing. When magma collects underneath a volcano, the ground above it sticks out. **Radar** instruments send radio signals from satellites to Earth's surface. The signals reflect, or bounce back, into space. Figuring out how long signals take to arrive is a way of measuring volcanoes. When a bump is detected, it takes a bit less time for the reflected signal to arrive. This shows there might be an eruption. **Global Positioning System (GPS)** instruments are so sophisticated that they can pick up tiny 0.04-inch- (1 mm) changes in a volcano's shape.

STUDY FROM THE AIR

Mount St. Helens erupted continuously between 2004 and 2008. To study the eruption, a helicopter flew over the volcano with two cameras: a thermal imaging camera and an ordinary camera. The pictures taken gave scientists important information about the eruption. It also allowed them to work at a safe distance from the volcano.

This image was taken a week after the ash and steam eruption from Mount St. Helens on March 8, 2005.

UP-CLOSE STUDY

Volcano experts may have to travel a long way to study volcanoes. They may have to visit dangerous sites to collect samples. Sometimes, they have to get very close to a volcano to measure signs that it may be about to erupt, such as escaping gas. In places where that is extremely dangerous, they use robots to help carry out the work.

Volcanologists who study active volcanoes close up wear protective suits with a metal coating that reflects the heat of the volcano.

Measuring Gases

If volcanologists think a volcano is active or may be about to erupt, they study its gases. These gases are released from the vent. They include water vapor, carbon dioxide, and sulfur dioxide. Magma contains dissolved gases. As it rises closer to the surface beneath a volcano vent, the mixture and amount of escaping gases can change. By collecting samples of these gases, volcanologists can spot any sudden change in the gases. That can warn them that the volcano will soon burst into life. The scientists may place instruments to measure those gases on the rim of a volcano crater or they may put them onto **drones**. The drones can be flown very close to the crater of a volcano.

Volcano Spiders

Robot spiders help volcanologists safely gather information about active volcanoes. The spiders are like portable labs. They carry a lot of different instruments, including equipment to measure vibrations and GPS to help find them. The spiders send **data** 24 hours a day to computers back in the office. They give continual updates about the activity of a volcano.

Robot spiders can be used to gather information in areas that are too dangerous for humans to visit. Once they have studied the sides of the volcano, they can even be lowered directly into it.

PLAYING SAFE

When learning to study volcanoes, volcanologists are taught safety measures. They learn first aid techniques and helicopter rescue procedures. They are also trained to wear gas masks and heat-resistant gear as needed.

Disaster Zone

Tungurahua, Ecuador

Tungurahua is a steep-sided 16,480-foot (5,023 m) volcano in the Andes Mountains in Ecuador, South America. The last big eruption from its summit crater was from 1916 to 1918. The explosions were so strong that pyroclastic and lava flows reached homes at the bottom of the volcano. Since October 1999, the volcano has been highly active and volcanologists have been watching it closely.

Studying the Volcano

Tungurahua is one of Ecuador's most active volcanoes. It is surrounded by several towns in which around 30,000 people live. It is very important to study an active volcano that has a high population close by because all of those people could be put in danger by a volcanic explosion. Volcanologists work permanently at Tungurahua's volcano observatory. This is a special building in which scientists monitor volcanic activity and keep in close contact with people in the area. They have trained volunteers from the community who take turns to watch the volcano 24 hours a day and keep scientists informed of any findings.

Tungurahua means "throat of fire" in the local Quechua language. The name perfectly describes the lava that can spew from this dangerous volcano.

Recent Eruptions

Since 1999, Tungurahua has had three major eruptions. An eruption in 2006 killed six people and destroyed more than 50 homes. Its heavy clouds of ash smothered crops and killed animals. The ash damaged roads and buildings. In 2014, thick ash spewed from the crater. A massive explosion of smoke and ash 5 miles (8 km) high followed. The volcano's lava flows blocked mountain roads. Today, the volcano is not quiet and frequently spits out some ash or lava. Perhaps it is warning everyone that it is ready to explode again.

Scientists have discovered that the Tungurahua volcano is steadily swelling, which is almost certainly a sign that new magma is building up below the surface.

UNPREDICTABLE VOLCANO

When Tungurahua had a minor lava eruption in 1999, volcanologists knew that three major eruptions in the last 300 years took place after a minor one. They acted quickly. They warned more than 15,000 people living near to the volcano to leave the area. Although the volcano kept burbling, it did not erupt. The residents returned home and some were annoyed that they had left for no reason. That reminds us how difficult it is for experts to be certain when a volcano will erupt and in what way—or if it will even erupt at all.

Chapter 5

SUPERVOLCANOES OF THE FUTURE

A supervolcano is a massive volcano with an eruption rating of VEI 8, which is the highest value on the VEI. The world's most recent super eruption took place about 26,500 years ago. It was the Oruanui eruption of New Zealand's Taupo supervolcano. It is possible that another supervolcano of a similar scale could erupt again in the future.

Crater Lake in Oregon is caldera 6 miles (10 km) wide and the deepest lake in the United States. It is the result of a supervolcano that erupted more than 7,000 years ago.

Spotting Supervolcanoes

Supervolcanoes do not look like volcanic mountains. They have very large, bowl-shaped craters called calderas, which form when the magma chamber of a supervolcano empties suddenly after a major eruption and the vent collapses inward. This leaves a large bowl shape in the ground. Calderas are much larger than normal craters. Parts of their sides may be missing. That's because land collapses unevenly during huge eruptions. Over time, giant calderas fill with water from rain or snow. The caldera created by the eruption of Taupo 26,500 years ago is now New Zealand's largest lake. It is called Lake Taupo.

Pyroclastic flows from supervolcanoes are believed to have changed the color of the sky in ancient times.

Monitoring Supervolcanoes

Scientists study caldera sites where supervolcanoes have erupted in the past to figure out if there are any patterns in their behavior. They want to find and study those patterns because, if they happened before an eruption in the past, they might happen again before a future eruption. If that is the case, it could give experts a chance to warn people. If they learn that the last two eruptions of a supervolcano happened 24,000 years apart, it might mean the next could also happen after 24,000 years. The problem is that even when patterns are discovered, patterns can change. Supervolcanoes could surprise us. They could act in a completely new way the next time they erupt.

Q

Why do you think it might be important to educate people about the dangers of volcanoes, and their unpredictability? How might that help when people are asked to evacuate because of a possible eruption?

Where in the World?

The smaller Hatepe eruption of New Zealand's Taupo supervolcano sent a plume of ash and gas 30 miles (50 km) into the atmosphere. It covered the entire country in ash to a depth of at least 0.4 inch (1 cm). Some parts of the country were buried in about 330 feet (100 m) of super-hot pyroclastic flow. It spread up to 56 miles (90 km) from the vent. Some people believe that ash from that eruption even caused red skies. The ancient Chinese and Romans reported seeing red skies at that time.

MAPPING VOLCANOES

Volcanologists gather a lot of information while studying volcanoes, such as how active a supervolcano is and when it might erupt in the future. They study the rocks in the area and information about current volcanic activity. They also study the way in which a certain volcano works.

Core Drilling

Core drilling is a way of finding out about a specific volcano or supervolcano. After a volcano erupts, material such as ash and lava settles. Over time, it is buried by other **deposits**, such as mud. Slowly, the extra material above presses on layers below, and that is how layers of rock form. Volcanologists use long, hollow steel pipes to drill for samples of these rock layers below Earth's surface. The rocks are collected and brought up to the surface in the pipes. They are called core samples. Clues to the past are found in the different layers. Ash, different types of rocks, fossils, and other material are all present in the layers. Those clues can provide information about when a volcano erupted. They can show what type of eruption it was, and how much material it produced.

Volcanologists will often work with specialist drillers to remove a rock core sample for studying.

Computer Modeling

Volcanologists take all the information they collect from the past and the present. Then they use it to make computer models or maps. For example, they use sensors to find seismic waves. Those can help them find large areas of magma in the upper crust below a supervolcano. They feed that information into a computer program, which copies the conditions beneath supervolcanoes. That helps the experts learn more about how magma, buried deep beneath Earth's surface, works its way upward and erupts onto the surface.

This seismometer is monitoring Mount Vesuvius in Italy so people can be warned of possible eruptions.

CLIMATE CHANGE DISASTER

Around 250 million years ago, the climate on Earth changed completely. The change killed 96 percent of life on the planet. Some scientists think giant ash clouds from a supervolcano blocked the sun, and that may have led to the huge loss of life. It is possible that a new supervolcanic eruption could produce enough ash to trigger a similar **climate change**, leading to another major disaster on Earth.

Disaster Zone

Yellowstone National Park

At the heart of the United States is a supervolcano that could blanket the country in a permanent winter. Yellowstone supervolcano has been asleep for more than 70,000 years. However, scientists say it is likely to erupt some day in the future.

If the supervolcano does erupt, the explosion would be 1,000 times more powerful than the Mount St. Helens eruption of 1980.

Monster Eruption

The volcano at Yellowstone National Park is in Wyoming and Montana. It sits on top of a huge **reserve** of molten rock and last erupted 640,000 years ago. Yellowstone is one of the biggest active volcanic areas in the world. An eruption from it could kill as many as 90,000 people—almost instantly. It would release a 10-foot (3 m) layer of molten ash that could stretch as far as 1,000 miles (1,610 km) from the park.

Changing Life

If there was an eruption, the volcano would release sulfuric gases. They would get into the atmosphere and mix with the planet's water vapor, causing a haze of gas to cover the United States. It would dim sunlight and cool temperatures, and would probably result in a year-long winter throughout the country. With the sharp drop in temperature and sunlight, crops would fail. That would lead to food shortages, both within the country and worldwide. It would be hard to travel because slippery, dangerous ash would cover highways. The ash cloud in the sky would make it so difficult to see that air travel would be impossible, too.

There are bison, beavers (below), bears, bobcats, moose, and wolves in Yellowstone National Park. All would be threatened by a disaster there.

TOP ATTRACTION

Many animals and plants live in Yellowstone National Park. The park has a very big tourist industry, too, and many park rangers work there. Guides give tours across its spectacular landscape, and hundreds of thousands of visitors come to see the park's incredible wildlife and **geysers** each year. It is hard to imagine such a beautiful site could be a deadly disaster zone.

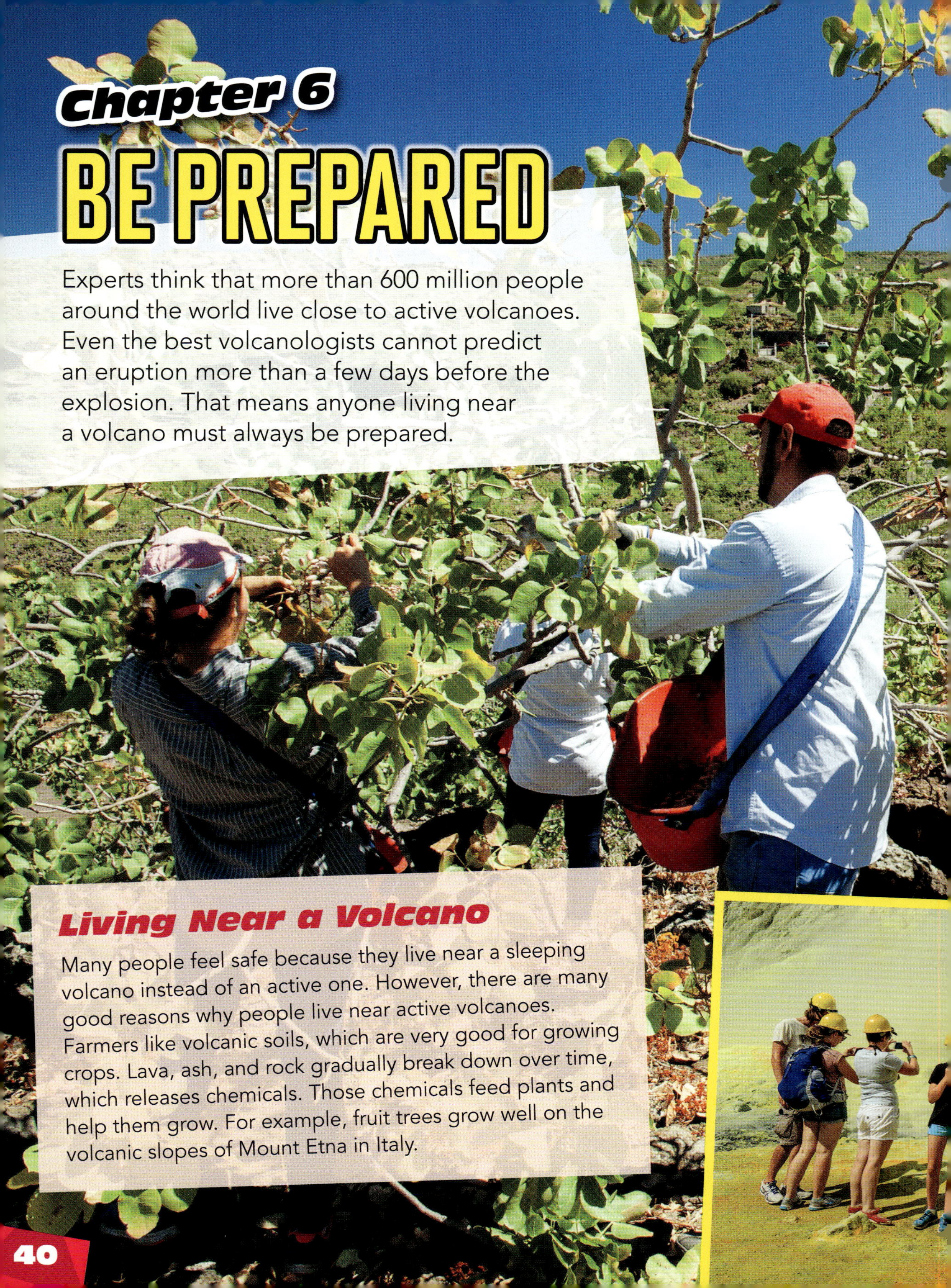

Chapter 6

BE PREPARED

Experts think that more than 600 million people around the world live close to active volcanoes. Even the best volcanologists cannot predict an eruption more than a few days before the explosion. That means anyone living near a volcano must always be prepared.

Living Near a Volcano

Many people feel safe because they live near a sleeping volcano instead of an active one. However, there are many good reasons why people live near active volcanoes. Farmers like volcanic soils, which are very good for growing crops. Lava, ash, and rock gradually break down over time, which releases chemicals. Those chemicals feed plants and help them grow. For example, fruit trees grow well on the volcanic slopes of Mount Etna in Italy.

Soil that has formed from lava tends to be fine and light, which is ideal for growing crops.

Mining for Minerals

Mines are often built near volcanoes because useful and precious minerals such as gold, silver, and copper are found beneath the ground where there have been eruptions. Those minerals come from magma. The super-hot water in magma dissolves metals inside it. When the magma cools, the metals are left in cracks nearer the surface. Miners then dig the minerals out of the ground.

Energy Supplies

Heat trapped beneath Earth's surface is called geothermal energy and can be used to produce electricity. Most of the electricity is produced by using steam that comes from natural sources, such as geysers, or from hot water brought by pipes from beneath Earth's surface. Countries such as Iceland and Costa Rica take the water from the magma in active volcanoes and use it to make electricity.

Tourists often visit active volcano sites because they are so exciting to see. However, they should keep at a safe distance from the volcano vent.

DIFFICULT DECISIONS

Volcanologists are very careful about predicting eruptions. Telling people to leave their homes, businesses, and jobs can be difficult for those people. If the volcano does not erupt, they may be angry, believing they have been troubled for no reason. Too many false alarms may mean that people lose confidence in the experts' advice. They may then refuse to leave when another eruption is **forecast**.

VOLCANO ALERT!

Anyone who lives near or visits a volcano region must understand the dangers and be aware of warnings. They should be prepared if the worst happens and a volcano warning is given.

Evacuation Plans

Experts prepare maps and models to show where lava or ash may fall in different possible eruptions. Those can help officials plan escape routes and decide which routes will be the quickest and safest for people to follow in an emergency. Officials sometimes put up signs that tell people which direction to take if they must leave. Families should learn these routes themselves. They must make sure everyone in a household knows what to do when a volcano warning is given.

Emergency and evacuation signs have symbols on them so they are easy to spot and understand.

Understanding Warnings

Volcano observatory scientists give alerts and warnings based on their study of information. Warnings may be given on the radio, television, and online. In many places, scientists use different alert levels. They use color codes to describe a volcano's activity. For example, green means no activity. Yellow means some activity has been noticed. Orange means there is an increase in volcanic activity. Red means an eruption is about to happen or has already started.

Disaster Kits

Anyone who lives near a volcano, or any other natural disaster zone, must have an emergency kit. It should include items such as a radio and flashlight with spare batteries. It should have goggles and breathing masks to keep out ash and dangerous gases. It should also include a first aid kit, dried food, bottled water, and clothes to last for three days for each person in the household. Having an emergency kit can be the difference between life and death if people have to leave with no warning.

It is best to keep an emergency kit in a backpack, which is easy to carry if people have to evacuate in a hurry or on foot.

READING THE SIGNS

When Mount Pinatubo erupted in a massive explosion in 1991, scientists were able to read the signs of the coming disaster. They told officials what was happening, and more than 75,000 people were evacuated before the eruption. That proves warnings can save lives if people listen to them.

SURVIVING SUPER ERUPTIONS

Scientists know they cannot stop volcanic eruptions. They also know there are around 1,550 possibly active volcanoes around the world, and around 20 supervolcanoes. While that might sound scary, science, technology, and being prepared can help people survive an eruption.

Surviving a Supervolcano Eruption

A supervolcano eruption is unlikely to happen without some warning. Before an eruption, the nearby ground would rise upward or bulge, and temperatures around the volcano would quickly go up. Sensors would tell that the mixture of gases in the air above the volcano was changing, and there would likely be a series of earthquakes. The magma would move rapidly up through the crust and ash might appear. To prepare for a supervolcano eruption, people would have to be evacuated from a large area far beyond the volcano itself. Then it would be a waiting game to see if ash clouds change the world's weather patterns, as has happened with supervolcanic eruptions in the past.

In the future, it is hoped that early warnings can be given to ensure people can evacuate to safety before a supervolcano erupts.

The mountain of Huascarán is all that is left of an **extinct** volcano. It has not erupted for at least 10,000 years and is not expected to erupt again.

Hope for the Future

The more volcanoes that scientists are able to observe and study, the better their forecasts will be. Improvements in the technology they use to study volcanoes are helping them better understand the warning signs. That will give people more time to get to safety. For example, scientists want to discover at what point the pressure under a volcano vent causes an eruption. They can then measure that pressure, which may help them give people a few days or even months warning before an eruption. In the meantime, organizations work together to learn from past disasters. They make sure their escape and rescue plans save as many people as possible.

DEATH OF A VOLCANO

Scientists believe volcanoes do not last forever. They say that after a few 100,000 years, volcanoes die. Some volcanoes only die after a final and very violent eruption. A volcano that has died and that experts are sure will never erupt again is described as extinct.

Glossary

active is erupting or is likely to erupt again in the future
atmosphere the blanket of gases surrounding Earth
avalanche a large mass of snow that falls down the side of a mountain
climate the weather in one place over a long period of time
climate change a change in the pattern of the world's weather
cone the hill formed as material from eruptions piles up around a vent
continents large masses of land, such as Europe and North America
crater the bowl-shaped hole at the top of a volcano
crops plants grown for food
data information collected for study
debris pieces from something that has been destroyed
deposits something, such as sediment, laid down by natural processes
dissolved disappeared when mixed with a liquid
drones small flying device with a camera
earthquakes violent shakings of Earth's surface
evacuate to move from danger in an organized way
extinct will not erupt again
flooding when normally dry land is covered in water
forces of nature powerful natural forces such as hurricanes and floods
forecast to predict a future event
geysers jets of very hot water and steam that shoot up from underground
glaciers slow-moving rivers of ice
Global Positioning System (GPS) a satellite system used to find directions and locations
landslide large areas of mud, soil, and rock that slide down a slope
molten melted
mountain ranges long groups of mountains
oceanic plate a tectonic plate at the bottom of an ocean
orbit the curved path through which objects in space move around a planet or star
particles very tiny pieces of material
plumes rising columns of smoke or ash
predict to tell what may happen in the future
preserved kept in perfect condition
pressure a pushing force
pumice a very lightweight volcanic rock
radar a system that uses radio waves to locate and measure objects at a distance
remote far away from towns and cities
reserve a supply of something that is available for use when it is needed
rift zone a large area where tectonic plates are moving away from each other
samples small bits of something
satellites human-made objects that orbit Earth
silica a substance found in sand, quartz, and flint, used to make glass
sulfur part of a smelly gas called sulfur dioxide that escapes from volcanoes
summit the peak, or top, of a mountain or volcano
superheated describes something that is heated past its boiling point

Find Out More

Books

Dodge Cummings, Judy. *Eruptions and Explosions: Real Tales of Violent Outbursts* (Mystery and Mayhem). Nomad Press, 2018.

Howell, Izzi. *Volcano Geo Facts.* Crabtree Publishing, 2018.

Nargi, Lela. *Absolute Expert: Volcanoes.* National Geographic Children's Books, 2018.

Nelson Maurer, Tracy. *The World's Worst Volcanic Eruptions* (World's Worst Natural Disasters). Capstone Press, 2019.

Websites

Discover the science behind supervolcanoes at:
www.usgs.gov/faqs/what-a-supervolcano?qt-news_science_products=0#qt-news_science_products

Learn more about how supervolcanoes work at:
https://science.howstuffworks.com/nature/natural-disasters/supervolcanoes.htm

For more information about volcanoes visit:
www.natgeokids.com/uk/discover/geography/physical-geography/volcano-facts

Publisher's note to educators and parents:

All the websites featured above have been carefully reviewed to ensure that they are suitable for students. However, many websites change often, and we cannot guarantee that a site's future contents will continue to meet our high standards of educational value. Please be advised that students should be closely monitored whenever they access the Internet.

Index

About the Author

Louise Spilsbury is an award-winning author who has written hundreds of wonderful books for children. She has written many books about natural disasters and has a healthy respect for these awesome forces of nature!